Lead Mining

An Introduction

Dalesman Books 1985

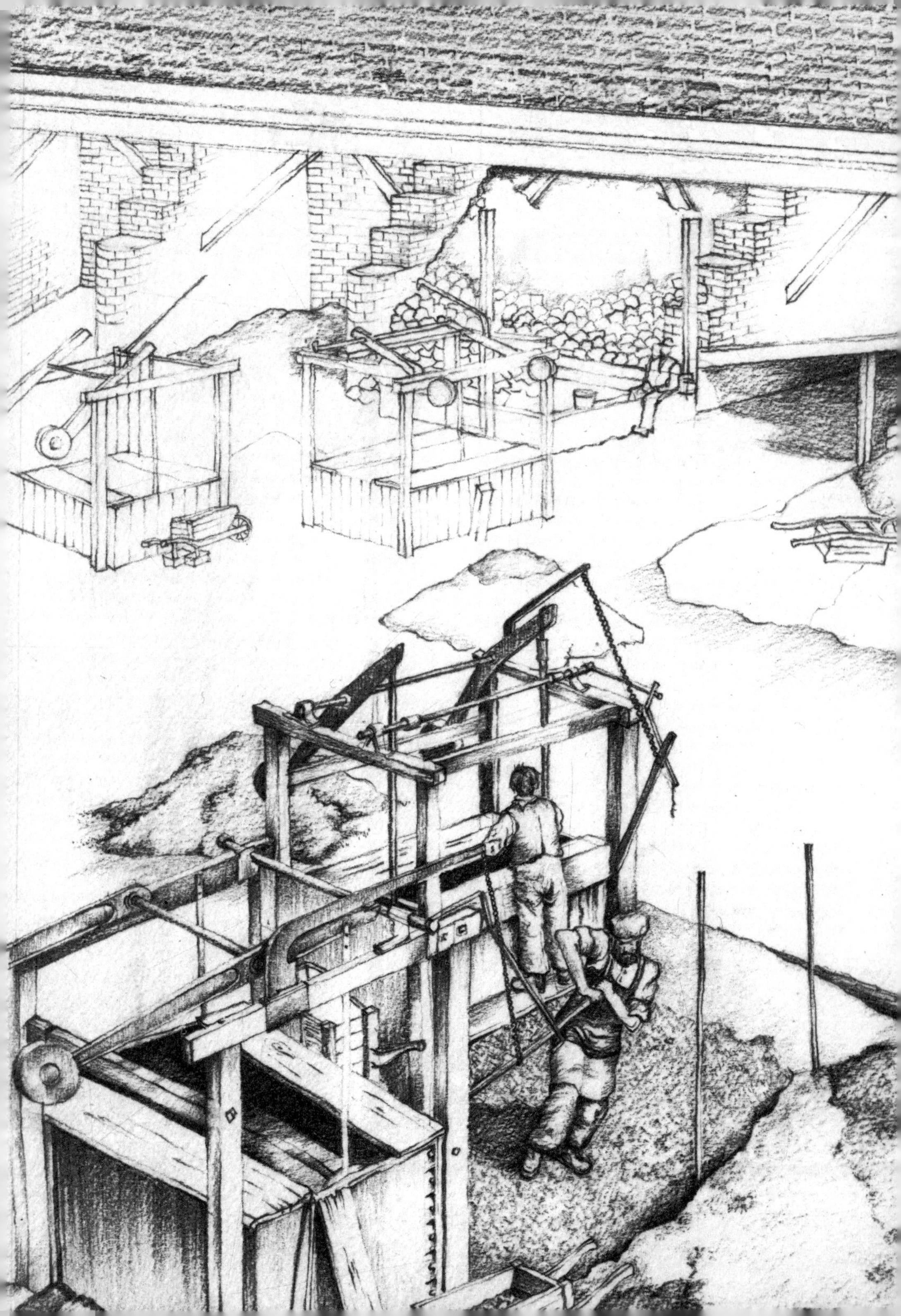

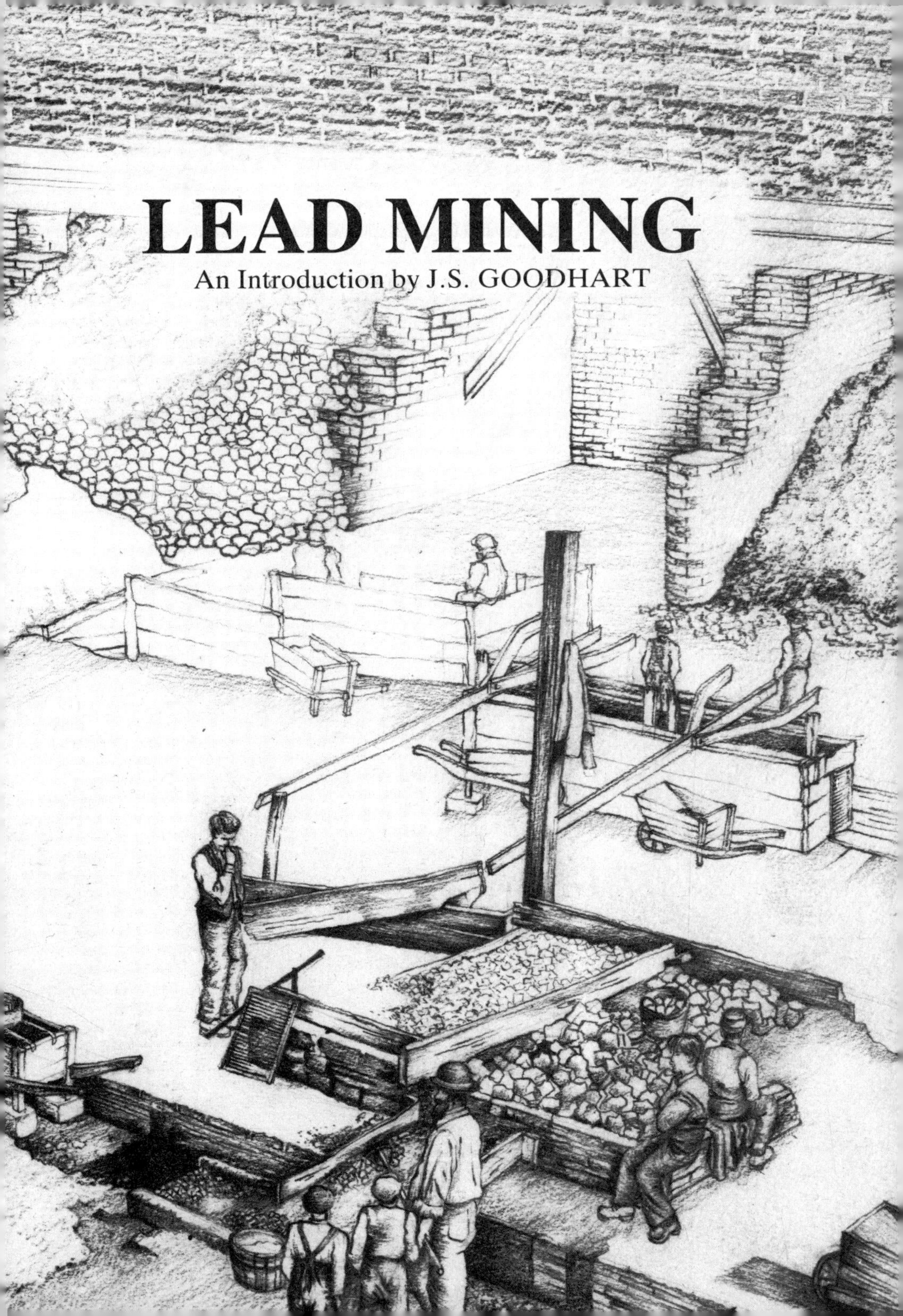

LEAD MINING
An Introduction by J.S. GOODHART

The Dalesman Publishing Company Ltd.,
Clapham, Lancaster, LA2 8AB.
First published 1985

ISBN: 0 85206 811 5

Dedication

To Vera, Kenneth and Stephen.

Printed by Fretwell & Brian Ltd.,
Healey Works, Goulbourne Street, Keighley, West Yorkshire.

Contents

Front cover photograph of the Killhope wheel by Geoffrey N. Wright

Safety Note

Discovering the countryside and the lead mining industry of the past can be very interesting but without sensible exploration it can be dangerous. The old mines and the buildings associated with them are generally in an unsafe condition. Since their abandonment they have been neglected for many decades.

My own exploring was conducted using maps and guides for hill walkers and my underground work was completed under the supervision of qualified outdoor instructors. Under no circumstances is it safe to enter a mine without an experienced guide. The lead mines are notorious for partially covered shafts where the timber covers have long since disintegrated and afford poor protection for those walking along tunnels. In the appendix I have recommended a kit list and included guidelines for obtaining suitable maps and books.

Acknowledgements

It is with thanks that I acknowledge the assistance given to me by Mrs. C.D. Leathley (illustrations); Mr. G.P. Young (drawing); Mrs. V. Young (typing); Mr. S.D. Goodhart and Mr. A. Crosier (photography) and by Beamish, North of England Open Air Museum for permission to study and use their photographs. I am also grateful for the advice of Mr. R.V. Turner (Alston Moor Historical Society); Mr. B. Rigby (geologist); Dr. T.M. Bell (Alston); Mrs. B. Reeve; Fitzwilliam Museum, Cambridge and Mr. P. Chandler of Sun House, Alston. I am particularly indebted to the pupils of Church High School, Newcastle upon Tyne, whose enthusiasm for the outdoors stimulated me to gather information which eventually led to the writing of this book.

Preface

ON seeing the bright circle of daylight at the end of the tunnel an immense feeling of relief swept over me. I was very glad to get out! Breathless and weary I collapsed onto the grassy bank, surrounded by the chattering of excited children. We had had a marvellous adventure, exploring the underground passages of an ancient Cumbrian lead mine. Suddenly the fascinating stories of the winning of lead became real to me. I had felt what the miner had felt. I had waded in the water where he had trekked, I had slipped in the same mud, battered my head on the same rocky protrusions and experienced the eerie, claustrophobic sensation of moving in a confined space.

Our party, being well led and clad in waterproofs and helmets, had suffered no physical adversity, save wet feet, a smattering of mud and a graze or two. Not for us the burning eyes and tight lungs of a tired miner. Underneath the fell we had found the air cool and clear, enjoyed the silence and perhaps for the first and last time in our lives sat in absolute darkness.

Being someone who is fond of Northern tradition and who enjoys dabbling in outdoor activities I was delighted to find on my own doorstep, so to speak, a new area of interest. I resolved to find out more by wandering the dales of the Wear, Tees, Allen and Swale.

Lead bracelet, thought to be middle bronze age, found at the excavation of Thesmi in Lebos.   *(Fitzwilliam Museum, Cambridge)*

The Attraction of Lead

LEAD is heavy, soft and grey and can be easily worked and moulded. As soon as people realise that something is useful they are prepared to pay for it by barter or with money. The lead miner came into existence because he could make a living or at least supplement his family income by extracting lead ore.

Man has searched for lead and used it for over two thousand years in our own country. Elsewhere there is evidence that the great civilisations used lead and this includes the Egyptians, Phoenicians, Chinese, Greeks and Romans. The early history of Britain is dictated by waves of different invaders. Bronze Age warriors came in search of land and cattle using bronze swords which had lead pummels on the end of their handles. The Romans used lead pellets in their slings, lead linings in their baths and they made lead coffins too.

The demand for lead grew over the ages until British mining reached a peak around the 1850s. By then lead was crucial for the manufacture of paint, pottery and glass. It was a basic material for roofing, ship-building, engineering and plumbing. The words 'plumber' and 'plumbing' come from the Latin word 'plumbum' which means lead.

At its peak the lead mining industry made the Dales some of the most valuable land in the country and was the closest rival to the coal industry for providing jobs. Unfortunately, by the end of the 19th century the whole era had ended and lead is now often replaced by zinc, plastics and other synthetics. Lead today still remains indispensible for acid storage, car batteries, X-ray shields, paint, sheaths for underground cables, petrol, soldering, printers' type and protection from radioactive materials.

LEAD IN THE GROUND

Hot gases and liquids are forced
up into surface rock cracks

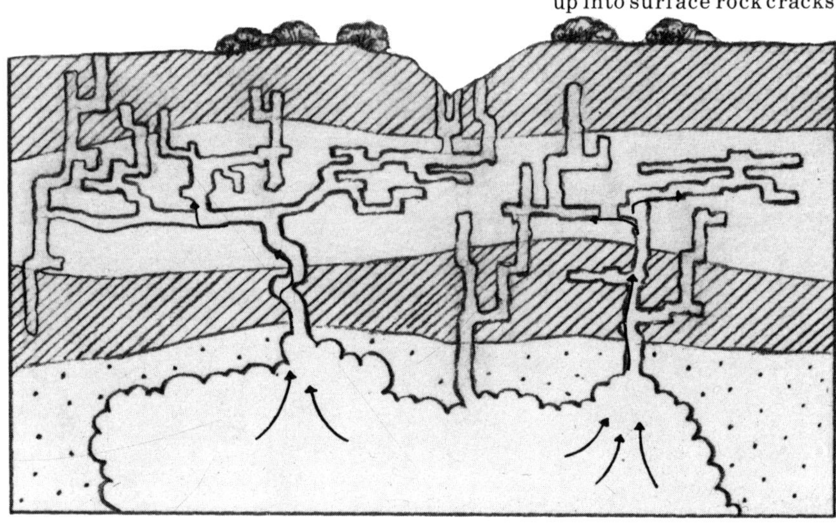

VEINS

Hot gases and liquids cool down in
surface rock cracks forming broken
and haphazard veins

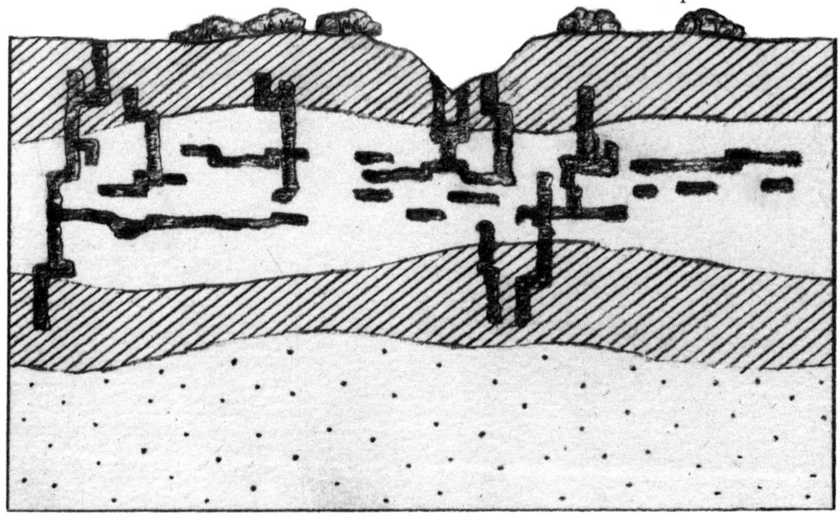

Lead in the Ground

BEFORE we look at lead mining it is important to understand how lead was formed, where it lies and how it lies. Millions of years ago hot gases and liquids from deep in the earth were forced into the surface rocks of the Pennines. (Figure 1). When these gases and liquids cooled they made some of the rocks and minerals which remain today. Lead is not formed purely, in the Dales it was and is still found within a mineral called **galena** or lead sulphide. Being a mineral which contains a metal, galena is referred to as an ore and has the properties of being very heavy and shiny when it is first cut.

Veins

MINERALS and ores are commonly found in **veins.** The galena rich veins of the northern Pennines occur in limestone and gritstone and are unusual, not only because they are long and thin and up to a few miles long, but because the majority lie almost vertically. Mining these **ribbon veins** can be extremely unproductive because they are broken and there is no guarantee that a vein will yield good quality **(high grade)** ore for the whole of its length. Sometimes the miner just kept digging in the hope that he would get a **lucky strike** and find that a dwindling vein started up again or the quality of the ore in the vein improved. It is not true to say that the miners depended entirely on luck for finding galena. They had special knowledge gained through experience. Firstly they knew that as the veins were vertical they could be followed up or down and secondly they found that veins were often parallel to one another. One of the best finds was when the miner came across a horizontal vein called a **flat.** Such a discovery was bound to cause great excitement because it was general knowledge that flats bore high grade ore and that extraction was relatively easy requiring less overhead pick work than extraction in a vertical vein.

A BELL PIT

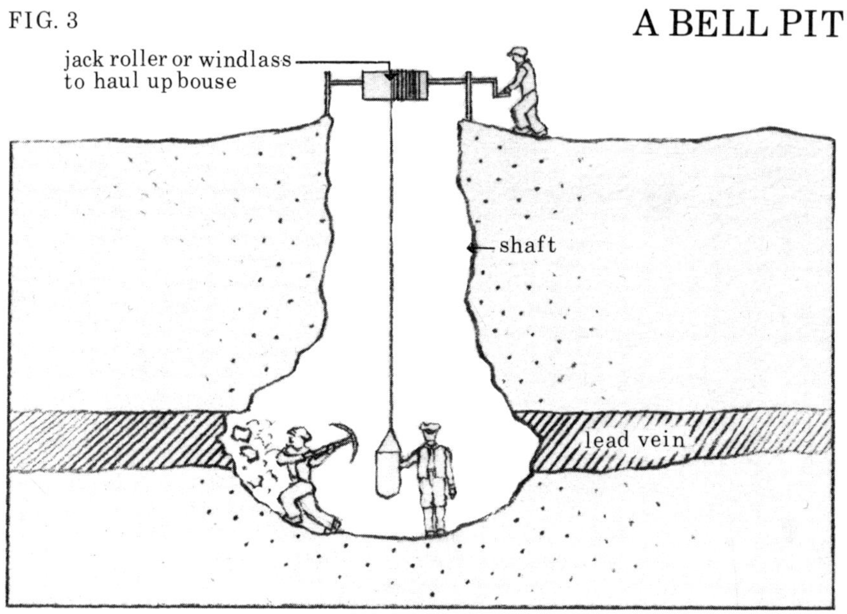

BELL PITS TODAY

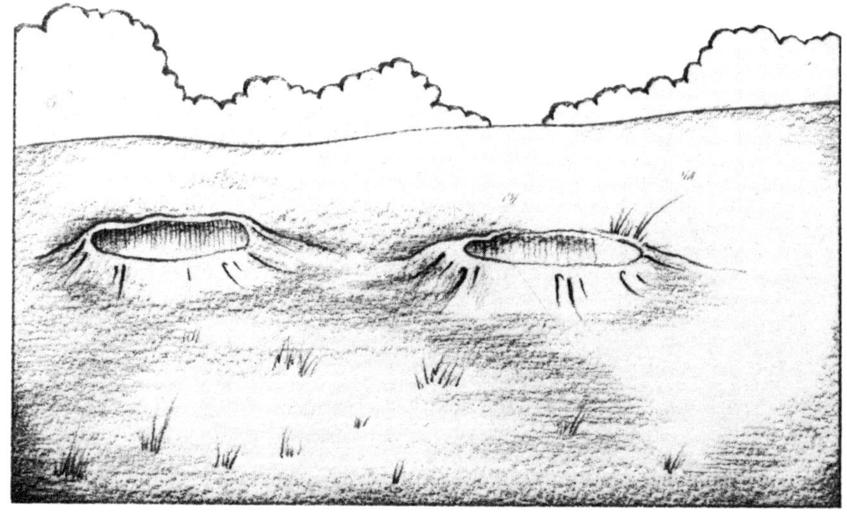

The Mining of Lead

Before the Roman invasion
Pieces of freshly broken galena would catch the eye of people who walked by a river bed. Where soil and rocks were exposed on the steep sides of river banks, weathering caused pieces to break off. These too would be spotted and collected.

Going a-shoading
Shoad ore is ore on the surface. In some areas after heavy rain, the locals would go a-shoading, searching for ore about the rivers and streams.

Opencast working
If veins were discovered then opencast work could begin, the aim being to extract the galena. There were four main types of opencast work; these were digging, firing, hushing and bell mines.

Digging
A vein was followed along the surface or into a valley side by digging.

Firing
Firing worked on the principle that heated rocks and minerals will expand and if this is followed by rapid cooling the rocks will crack and split. A roaring fire was built against the rock and this burnt out early in the evening when the temperature started to fall. Cooling could be done by dousing the rock with vinegar and water if the miners did not want to depend on the night being cool. Water on the hot rock would turn to steam and the evaporation speeded up the cooling. Once loosened, the miners would set to their work using hammered-in wedges and then picks and shovels. In the open air firing was free from the dangers it held when used under the surface. In a confined space the miners suffered the fumes which rose from the heated minerals.

Bell Pits
BELL PITS were named after their bell-like shape. Early man also used this method in the mining of flint. A shaft of about twenty-five to thirty feet was dug down, then the bottom was opened out until the earth walls became too dangerous and collapse was imminent. (Figure 4). The **bouse** (which was made up of unwashed ore and unwashed waste minerals termed **gangue**) and

HUSHING

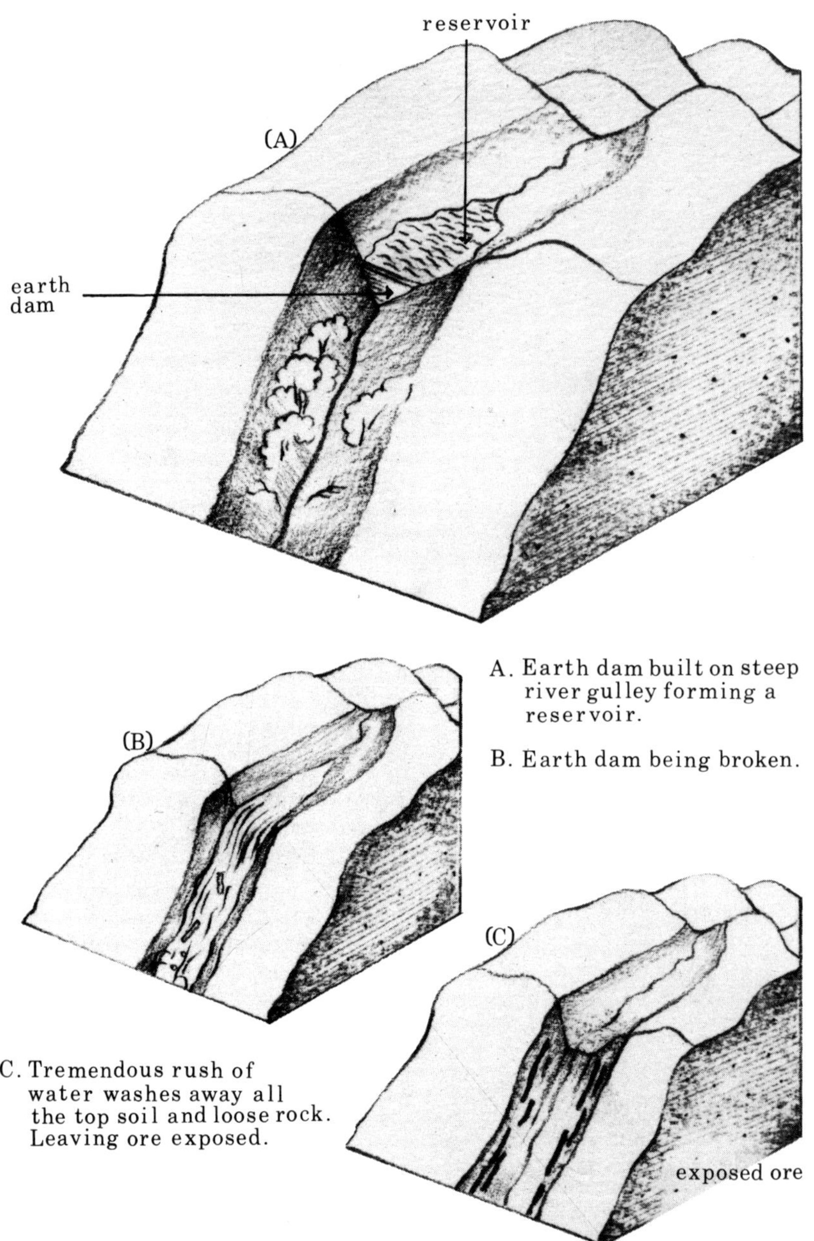

A. Earth dam built on steep river gulley forming a reservoir.

B. Earth dam being broken.

C. Tremendous rush of water washes away all the top soil and loose rock. Leaving ore exposed.

the **deads** (waste rock) were either carried or hauled out. The hauling or carrying was done by the **miner's mate.** This could have been another man but it is likely that some of the miner's children were given this labourious task. The miner's mate also had to scoop and carry out water in bucket loads. Water was a danger and a hindrance because it softened the shaft sides and speeded up the disintegration of the walls.

Bell Pits Today

Numerous old bell pits can be seen around the old lead mining areas. On Ordnance Survey maps they are often marked with a tiny circle and labelled old shafts. Some have collapsed and have a pool in the bottom and many are half filled in by farmers or fenced off to keep straying livestock and walkers out. Standing by other bell pits, showing the standard warning mark of countrymen, are large piles of stones which are the traditional cairns. All of the old pits have one obvious characteristic, a circular mound where bouse and deads were dumped around the edge of the shaft mouth. Most of this rubble is now grassed over.

If it was possible to view the mounds from a helicopter they would be seen following straight lines across the landscape. They were arranged this way because it was vital to dig the shafts along the line which the lead vein took on the surface.

Hushing

The Romans had lead mines around Rio Tinto which is on the north-west coast of Spain. Their lead pipes have been found in both Pompeii and Britain. Hushing was a method of exposing ore which is said to have been used as far back as Roman times. An earth dam was built at the head of a river gulley or on a fell side where there was a chance of finding ore. A great lake or reservoir was then formed behind the dam. (Figure 5). Later, when the lake was big enough the dam would be burst and the tremendous rush of water would wash away all the topsoil and loose rock. The gulley or fell could then be worked using suitable open cast methods.

The name **hush** may come from the hushing sound water makes as it rushes down a valley. Hushes are often named after people and in Weardale you can still find Milburn's Hush, Peter's Hush and Adam's Hush.

INSIDE A NEW MINE

FIG. 6

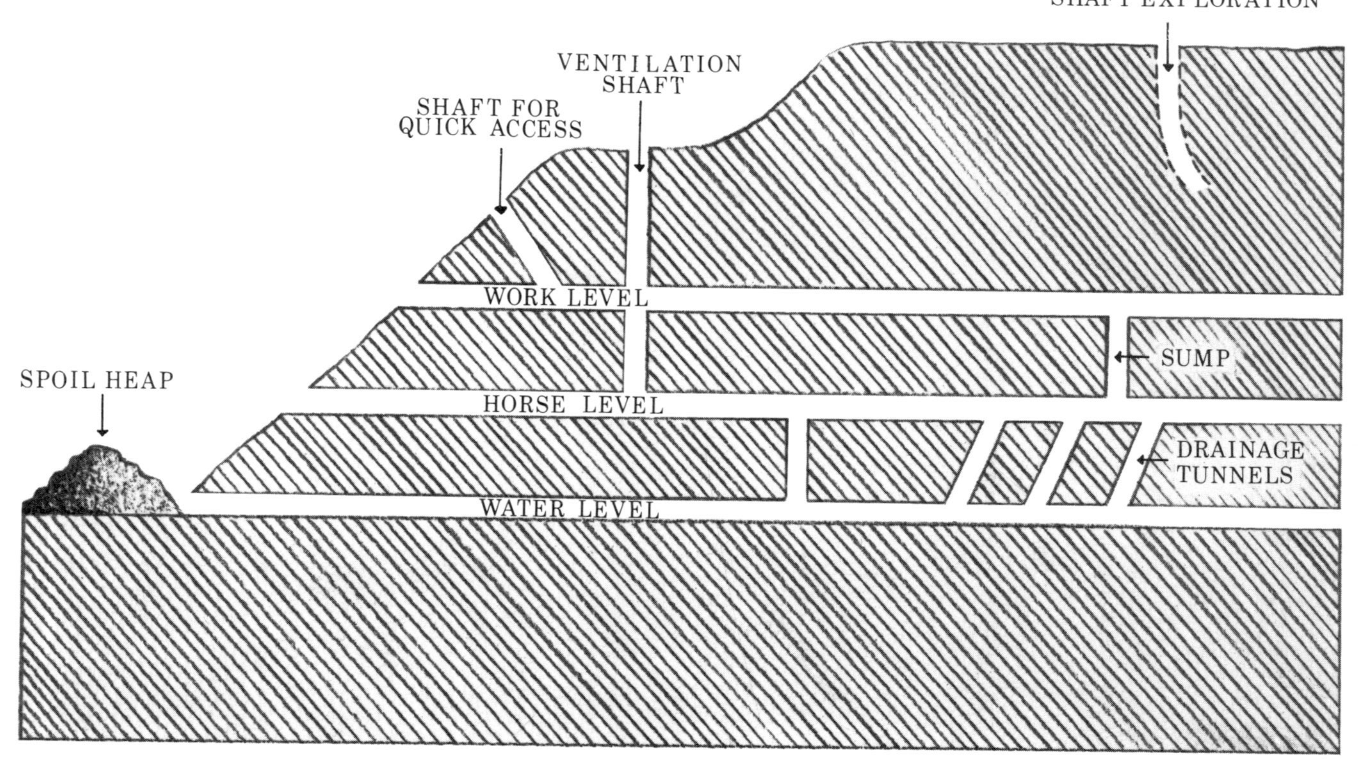

New improved Company Mines

THE new type of mine was made up of horizontal tunnels driven into the fell. (figure 6). These tunnels were called **levels.** Advantages of working in these mines were numerous. Access was good, the miners could walk in and out. Load carrying was easier because rails were laid so that the men or ponies could bring out the bouse and deads in trolleys. Improved drainage meant quicker work was undertaken and ventilation overcame the problem of impure air to some extent.

Inside a new Mine

The work level: Ore was removed from here.

The horse level: Ponies or men would drag out the bouse and deads in large tubs which ran along wooden rails.

The water level: Water was re-routed or collected. It was taken from the work level to improve working conditions.

Sumps: These short shafts connected one level with another. Bouse and deads were dropped down here to be hauled away.

Shafts: Shafts were used for the exploration of new veins and to improve access. Occasionally lead was removed from here.

If bouse and deads were to be removed from a deep vertical shaft then a large jack roller was used, another name for it was a windlass. (Figure 4). If the shaft was too deep then a **whimgin,** a large jack roller on its side, was pulled by a horse. A **gin-gan** worked on the same principle, it was a horse powered machine for grinding corn which farmers used at that time. A gin-gan can be seen at Home Farm in Beamish (The North of England Open Air Museum).

Why the small miner disappeared

THE large companies wiped out the small miner during the 18th century. The London Lead Company, also known as the Quaker Company, was based in the boom town of Middleton in Teesdale and was a good example of a company which helped ruin the independent miner. The reasons for the

dominance of the large companies are set out in the table of comparisons below.

<table>
<tr><td>The Small Miner</td><td>The Company</td></tr>
<tr><td>Exploration and extraction could not be undertaken at the same time.</td><td>Finance was provided for exploration while other valuable ore was still being worked.</td></tr>
<tr><td>Prediction of the quality and the quantity of ore was impossible so he often mined poor (low grade) ore.</td><td>Profitable sites (like Nenthead in Cumbria) were pinpointed and worked.</td></tr>
<tr><td>Conditions of work were rough and dangerous.</td><td>Mine drainage and ventilation were improved.</td></tr>
<tr><td>Pay was very poor.</td><td>Pay was low but slightly better than the pay of a freelance miner.</td></tr>
<tr><td>Poaching was almost the only way to supplement the family diet if times were difficult.</td><td>Small holdings were often provided.</td></tr>
<tr><td>There was little chance of any education for the miner or his family.</td><td>School houses and chapels were built.</td></tr>
<tr><td>Living conditions were poor.</td><td>Community buildings like wash houses and reading rooms were built.</td></tr>
<tr><td>There was never enough money to buy machinery or to improve haulage.</td><td>Hydraulic power was developed for dressing the ore. Teams of ponies, with a man to control them, were hired to transport ore.</td></tr>
</table>

In summary, the companies could offer safer working conditions, better pay and some social and welfare considerations including a place to sleep at the site of the mine. (lodging shop). By modern standards even the improved conditions provided by the company would be considered inhumane.

Garrigill: tunnel entrance to a drainage level, now closed up by the council. *(S.D. Goodhart)*

A PICKMAN

1. FELT HAT
   Providing warmth,
   protection from
   dust
2. SCARF
   Providing warmth,
   a mask against dust,
   sweat absorber

3. HARD-WEARING
   SHIRT
   Loose sleeves for
   easy movement
4. LEATHER WAIST-
   COAT
   Draft and water
   proof

5. BAGGY TROUSERS
   For easy movement
6. HEAVY CLOGS
   Hard-wearing
   leather soles

BOUSETEAMS

Tools and Trades Underground

Deadmen or Sappers

These men drove the levels and the shafts. As equipment they would have hammers, sharp iron spikes for boring, gunpowder, charges, tapes, matches, rammers, powder cases, candles and lumps of clay to stand their candles in. Part of their job was to bore holes in the rock, ram in the gunpowder and affix charges and tapes. The men would then take cover, put their scarf up as a mask against the dust, light the tape and wait for the explosion. The explosion would crack the rock so that it could be taken out to deepen the tunnel.

The teams of deadmen would **strike a bargain** (make a deal) with the lead company over how much pay was given. Pay depended on how much work was done and how hard the rock was. Men who worked with explosives often bore the scars of scorching on their faces and the fumes and dust they encountered eventually gave them lung disease.

Pickmen

Pickmen worked the veins of lead ore and are the men we think about as we imagine a miner at work. (Figure 7). Their tools were picks, hammers, strong oak wedges to crack the rock, shovels and clay, candles and matches to give some light. Later miners used acetylene lamps fixed to the peak of their caps and these had to be filled daily. Their shovels were not like the traditional shovel for clearing earth and snow. The edges were not curled and this ensured that the shovel could be swept to the left or right to scrape up all the bits of rock.

Like deadmen the pickmen worked in teams or **partnerships** as they were called. It was important to keep the bouse of each team separate because pay depended on how much lead their bouse yielded. The bouse was kept in walled bays called **bouseteads** or **bouseteams** (Figure 8) and pay was by the **bing.** A bing was 407kg or 8cwt.

Miners at work. Drilling to take explosives in mineral vein (shot firing).   *(Beamish Museum)*

Hazards

Explosion
Stone dust and fumes entered the lungs causing lung disease and the coughing and spluttering of 'black spit' which characterised the underground worker. Burns and injury from flying rock were also possible.

Choke Damp
This was the name given to the air when it had such a great proportion of Carbon Dioxide in it that the miner gasped for breath and had to go up to the surface for fresh air. The Choke Damp gave the miner a headache and could lead to suffocation if there was no opportunity to get fresh air.

Lead Poisoning
Lead poisoning made the miner feel tired and faint. Other symptoms were dizziness, violent headache, sickness and acute stomach and bowel pain that is known as 'lead colic'. Through smiling faces, the black line at the base of the gums was a sign of anaemia and severe poisoning was marked by lead palsy, a paralysis of the hands where the wrists droop.

Cold and wet
Rheumatism and arthritis were aggravated by the cold and wet. Pneumonia developed from other breathing ailments and skin eruptions would not heal.

Tiredness
Walking to work and the arduous nature of the labouring made the miners physically tired. Bad temper, depression and general tiredness made other medical problems worse and were the root cause of some accidents.

Flooding
Drowning was possible.

Accident
All kinds of injury occurred. Some were regular and minor like scrapes and bruising but a number of accidents were fatal.

Working life
Company records would lead us to believe that the miners died through misfortune — they could not have possibly have died through neglect! It is disturbing to note that the health hazards faced by the miner at home and at work gave him a life expectancy of only 47 years, when the national average was 62 years for a man.

Allenheads Yard: Dressing ore. *(Beamish Museum)*

Discomforts and Suffering

STUMBLING and falling were two of the most common incidents which resulted in cuts and bruising. Everyone would suffer muscular strains because of the heavy work and bending in cramped conditions led to backache. Any bony area like the knees, elbows, back and knuckles were regularly skinned and any sores were irritated by chemicals in the water.

Haulage arrangements caused many accidents, horses would kick and trolleys could knock people over or run over their feet. If any bones were broken and were not set properly then that person would end up with a bone that was permanently bent. Many workers had stooped and twisted walks and bent arms, fingers and toes.

If children worked underground they suffered too. Many of them did not receive regular meals which provided the nourishment for them to grow. It was mostly boys who were allowed to help with underground tasks like pushing trolleys and leading horses. Above the ground children might help with processing the ore. Spare the rod and spoil the child was the maxim of the day and little sympathy would have been wasted on tired or crying children. It was in the nearby coalmines of Northumberland, Durham and Yorkshire where more cruelty to children occurred. Children there were left to open and close ventilation doors. Being alone for hours with no rest, no friends to talk to or decent food they often fell asleep and thus were instrumental in causing many mining tragedies. It is not surprising then, that for children who were lazy or tired the punishment was cuffing and beating.

Drag your feet to work my son,
Stumble in the half light.
Earn an honest penny,
Don't fall asleep afore night.

The exploitation of children and the conditions in which they live and work is always a subject which evokes an emotional reaction. We can learn a great deal about the pitiful childhood of child miners, washers and dressers by reading the essays collected by Dr. Mitchell in 1842 for the Children's Employment Commission. Extracts from this report are presented below:

Stephen Collingwood
I am 15. I work at washing lead-ore; I come at seven and leave at six . . . The boys' play are marbles and jack-ball; we play cricket and football. I often go fishing in the river and catch trout; I never go bird-nesting. I go to church on Sundays sometimes twice . . . We have wooden

soles to our shoes to protect our feet from the wet, and iron about the sides and heels to make them strong. The mine is warm but sometimes very bad for air; we can hardly breathe . . . Very few work till 50 years of age.

Joseph Fleming
I am 14. I strike with a hammer and bring down the mine . . . after bait we work till eight hours work is done. I go to bed at nine. I come home and put on other clothes; I wash by face and neck; on Saturdays I wash down to my navel; after washing we have a few bits of potato and mutton.

Joseph Collingwood
The miner's breakfast is usually crowdy of oatmeal. Some takes a little supper but some not. We can work at lead mines . . . our mines do not excite the appetite like the coal mines. We do not eat so good victuals, nor near so much, nor do we drink much beer, because we can not afford such things. Boys have a better appetite than the men, and take crowdy twice a day; they may get a little milk or beer to their crowdy. The London Lead Company in '39 and 1840 imported a great deal of rye-corn, and had it ground and sold to the workpeople. The bread did not agree with many, and those who could possibly do without it gave it to the pigs; the medical men disapproved of it as food . . . the miners have very little fresh meat . . . There are far more widows than there are men who have survived their wives.

Ralph Elliott
I went into the lead trade to wash ore at 10 years; there are very few under 10 and none under nine. There is a shed in case it rains, and their is a fire in it; there is a shed at all the London Lead Company's works but it is not usual at other places. We work, if possible, not withstanding rain, but if it rain very much we go home; we change our clothes when wet . . . sometimes I sit at home and read the Bible; I never read any other book; I can say the Lord's prayer and Creed; I say the Lord's Prayer sometimes but not regularly.

George Sanderson
I am 15 years of age. I work at washing lead-ore in the summer, and in the mines in winter. I get 4 shillings, sometimes 5 shillings and I get about the same in the mines. I have read the Bible and the History of England and nothing else.

John Rain
There are about 30 men in our mine but very few boys. Some days the air comes well into the lower level, some days not; and what between the powder-reek, and the want of fresh air, we have some great difficulty of breathing. I am not aware of any difference of mine between summer and winter, but when it is wet and dull the air is bad. There are four of us in our partnership, and we fire off six or eight shots a day.

Processing the Lead Ore

THE galena was processed by men, women and boys until there were fewer jobs because the price of lead was too low for it to be profitable for the company. The population had increased as the mines attracted a workforce and there was competition for jobs. The companies could choose the strongest workers who were usually the men or the cheapest workers who were the boys. The first stage in the process of refining lead was to **dress** the ore and the second stage was to send the dressed ore for **smelting.**

Dressing

Knocking
The bouse was rested on a flat knockstone and the boys and women wielded their large hammers to crack the bouse thus separating galena from deads.

Picking
The galena was picked out and separated from the deads.

Crushing
Further hammering smashed the galena into pea-sized pieces. Machines eventually did this job being powered by huge waterwheels like the Killhope wheel. Nenthead, a Cumbrian village not far from Weardale's Killhope site, had the second biggest wheel in the world for a time. The wheels could be put out of action by extremes in the weather. Ice restricted the water flow as did dry weather and without moving water the wheels remained motionless.

Washing
The crushed rock was sieved and the smallest pieces of dead rock removed. When this was done by hand the workers used a sieve and a tub of cold water. Modern processes add oil to the water at this stage because the galena becomes attached to the oil which rises to the surface. The oil and galena are then skimmed off leaving the waste rock below.

Waste
Waste was left in **spoil heaps.** By looking around spoil heaps and cracking open their rocks you can still find many interesting rocks and minerals.

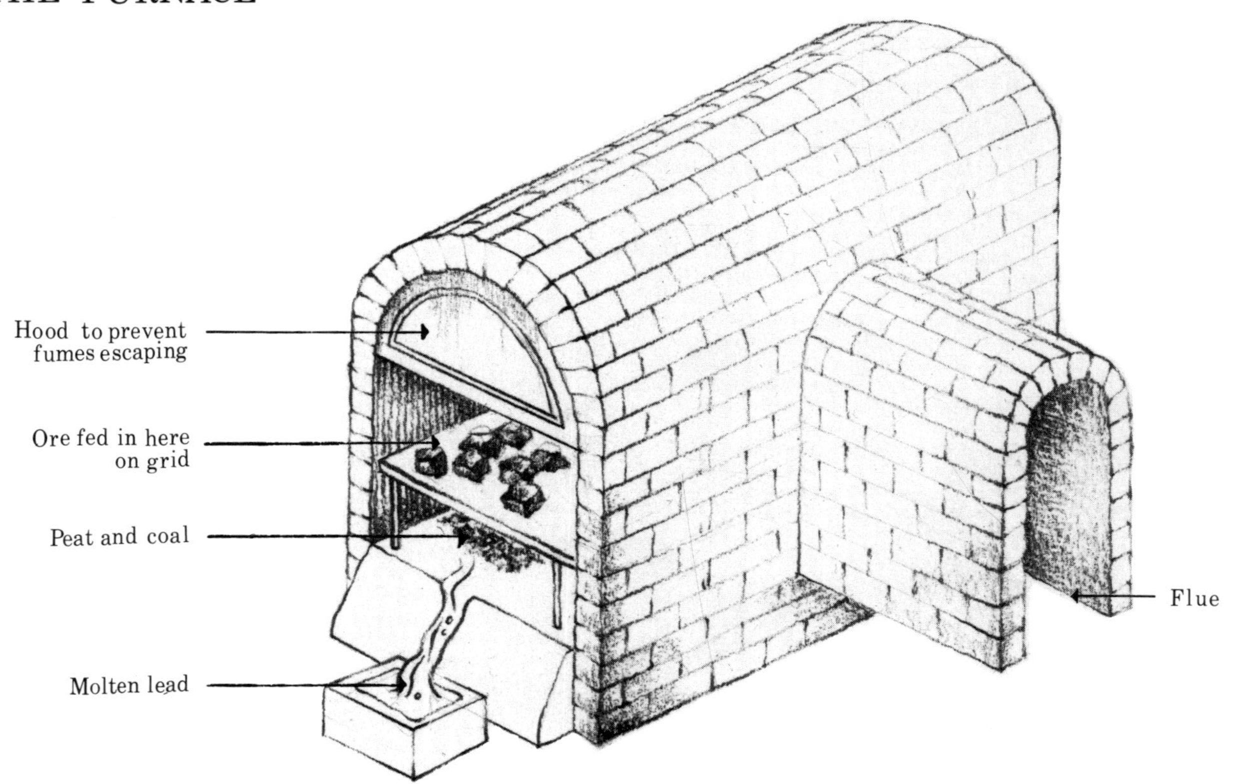

Hood to prevent
fumes escaping
Ore fed in here
on grid
Peat and coal
Molten lead
Flue

Smelting

Before substantial roads were built the dressed ore was taken over the fells to the smelting place where it was to be refined. Packhorses were used for this journey until horses could pull carts along the newly built company roads. Smelting was carried out near a good supply of peat, coal or wood which had to be kept dry. Although roofless, some of the fuel sheds are still standing, for instance at Gunnerside Gill in Swaledale where sturdy peat drying sheds now give the sheep some makeshift shelter. As the ore was heavy and transport had to be paid for, the closer the smelt was to the mining site the better.

In early times smelting was simply done over an open fire on a fell or hill but as the smelting process was developed furnaces with long, thin hearths were used. (Figure 9). In a furnace the temperature could be controlled, the smelter using long pokers to reach into the furnace and huge bellows for fanning the flames.

By the peak of the lead mining boom lead mills, which were furnaces with tall chimneys, were used for smelting. The chimney caused a strong updraught which made burning more effective and carried away the dangerous fumes. The flues which led from the furnaces to the chimney were up to two and a half miles long! (Allen Mill near Allendale Town had a flue system this length). The advantage of such long flues was that it ensured that the lethal fumes were taken away from the workers, the local community and the fell livestock. Regular flue cleaning was carried out by small boys about once a year. The boys were assigned to the task of scraping off the **fume** which was caked to the sides of the flue. It was a dangerous job, some of the flues ran under the fell and death as a result of lead poisoning was almost inevitable. The fume was valuable and could be smelted for extra lead and extra profit.

Today, livestock is at risk from grass which is contaminated with lead derivatives from old spoil heaps. The rain may wash the waste onto the grass. Animals can also become contaminated when they breathe the fumes of heavy traffic; these fumes contain lead which was originally in the vehicles' petrol. The danger to humans is that to eat offal from contaminated beasts would make us all ill, but this problem is almost unheard of in Great Britain.

The Smelting Process

Modern methods of refining lead are based on the same techniques that were developed by smelters long ago. There are two main stages, these are **roasting** and **reducing.**

Roasting

The galena (Lead sulphide—PbS) is heated to burn off the sulphur. In modern smelting, coke is added because it readily combines with the sulphur to form a fused sinter or clinker which is easily removed. In burning, the lead combines with oxygen in the air to form the compound lead oxide. (PbO).

Reducing
The lead oxide is burnt to remove the oxygen and leave molten lead. (Pb). This lead has impurities in it but if it is kept just above its melting point in a large container then stirring will bring the **dross** (the impurities) to the surface. The molten lead used to be stirred in a **drossing kettle** and the surface was skimmed to remove the dross. Ingots of **pig-lead** could then be moulded.

Modern refining
Pig-lead is lead in a crude form direct from a smelting furnace. Modern refineries force air through the molten lead at the drossing stage and this has a similar effect to stirring because it helps to bring the impurities to the top. The smelting process is taken a step further by removing silver, zinc, bismuth and other impurities. To do this, chemicals and electrolysis are used.

Rookhope Smelt Mill.   *(Beamish Museum)*

Transport: This is the Age of the Rein

THE transportation of ore was originally the job of the small packhorse. In Allendale the breed used were **Galloways** but the local people nicknamed them in an endearing way calling them **'Galls'.** The alternative name for the horses was the **jagger** and some miners kept one on the land they owned or the land which the lead company had provided.

The horses were only twelve to fourteen hands high and a hand being four inches meant that their shoulder height was just over four feet. Even so, if they worked pulling tubs in the mine they had leather backskins to prevent their backs from being injured in the low tunnels.

Above ground the horses carried ore to the smelt mills by the quickest route. This meant taking short cuts across the fells which became known as **carrier ways.** (see Appendix). The main highways ran along the valleys and connected the villages so they were not always the best routes for taking ore to its destination. The horses made their way across the fells, tethered in groups of twenty to twenty-five and led by the **raker,** a reliable and belled horse who set the pace for the others. If a team of horses passed through a village they were greeted by a throng of excited children. There must have been some disappointment however to find that all of the animals were muzzled, not because they were aggressive but to prevent them eating lead contaminated grass or drinking lead contaminated water.

Changes in transport

The coming of the railway was too late for it to have been used extensively for the transport of ore and refined lead. Much of the lead produced was sent to Newcastle or Stockton for distribution, the latter being famous for its railway. Changes in transport occurred slowly from packhorse to cart to rail, horsepower having had most impact overall.

Lodging Shop.   *(Beamish Museum)*

The Caring Company

THE large lead companies had the financial resources to improve working conditions and this influenced the health of the miners. Chest complaints were rife and a combination of mining and home life caused illnesses which led to seasoned miners taking to their beds and dying after six weeks. It was in 1864 when this was reported to the Mine Commission but in the early 19th century some relief to the problem had begun through **watercoursing.** Some mines had a small stream passed through them and as the moving water created a draught the air was changed and any dangerous gases, fumes and dust were carried away. Extra tunnels were bored which were solely for the purpose of letting in the outside air to the deep levels. **(ventilation shafts).**

To relieve suffering and give some relaxation for the workers it was common for the miners to take smallholdings or cottages with large gardens. The London Lead Company and other companies used this ruse to help the miner get some fresh air and grow food; this meant that the miner would be fitter for working and increase profit levels. The miner for his part was grateful for an opportunity to keep livestock and grow food for the family. They may have kept a few sheep, a cow, a pig, some hens and possibly a local type of pony or horse.

The London Lead Company also increased morale by building schools, reading rooms and small cottages in addition to bringing a water supply and expanding the road system. Even though the lead companies improved the roads they were very poor compared to those of today. If it was too far to walk home and return by the next day then the miner would stay in the **lodging shop.** These shops were commonly used in Teesdale where the mines were often some distance from the local villages. Elsewhere the miner may have been able to go home at the end of the day, but it was not always possible.

Lodging Shops

Places of past lodging are marked on local Ordnance Survey maps but it is debatable whether they really improved the miners' welfare because they were so poorly kept. Clues to their whereabouts can be found in placenames. In Allendale we find Shop House at Sinderhope and Cow Shop near Whitfield. The recently renovated lodging shop at the Killhope Wheel site in Upper Weardale is a fine example but it is now far cleaner and habitable than it would have been in former years.

In a lodging shop the miner had shelter, ate his food, shared some form of lighting and slept in a bed for the night. The lodging shops were important because they allowed workers to escape the harsh Pennine weather which could be cold, wet and windy even in the summer months. Although built with good intention, the lodging shops were often as unhealthy as the miner's place of work. In 1842, in a report to the Children's Employment Commission, lodging shops were described as being 'not fit for a swine to live in.'

The reasons for this were that the beds were often infested with bugs, there was poor sanitation and little cleaning was done because the men were too tired and considered cleaning a womanly chore. Dirt, scraps of food and peelings were left to go mouldy in the pools of water on the floor and this encouraged both rats and mice. The miners had hardly any spare clothing, in fact they may not have owned other work clothes and there were no facilities for drying save hanging clothes from the rafters or ceiling. Mining was a dirty, sweaty job and as personal hygiene was not practised the damp morning stench of the lodging shop must have been unbearable. There was severe overcrowding with up to twenty people sharing one room with dimensions of 13ft. × 10ft.

It is not surprising then that these places were dubious as places of rest, the miner having to weigh up the advantages of avoiding the trek home with the disadvantages of the environment of the lodging shop.

Character, Comradeship and Community — in work and in leisure

THE miners were dedicated workmen and real men of mettle. Their steady application to their difficult job was a necessity to earn extra pay and their attitude was enhanced by the fact that most miners were strongly religious. In church and chapel there was the constant reinforcement of the 'work ethic', not only in connection with their employment but with the fruitful use of their leisure time. It is small wonder then that the lead companies were keen to provide churches and chapels!

In their spare time the miners could be found following many creative hobbies such as carving, knitting, darning or embroidery. Although we now associate some of these activities with women it is clear that monetary and clothing needs increased the miner's versatility! The creative pursuits may have earned a few extra shillings which were vital to survival if the price of lead dropped. Other activites such as gardening or animal keeping were carried out in the fresh air and were particularly therapeutic for the miner who was in poor health.

In the face of such a hard life it was important to have close supporting friends. Each man had his favourite work mate of **shoulder fellow** and they would keep each other's spirits up when religious solace seemed far away. At work, singing popular or dialect songs and reciting poetry were ways of increasing morale. In hard times the miners would set out across the fells to poach the 'bonny moor hen', the red grouse. It was a rare occasion to be apprehended because the moors were so large and the miners knew the fells so well.

The fact that poetry and singing were appreciated is a reminder of the high regard which the miners held for literature, knowledge and education. Last century a Mr. Foster described the lead miners as being 'remarkably intelligent and well educated'. Such was the interest in reading that the lead companies not only built reading rooms but provided heat and light for them too. In times when there was no work at the mine and the winter weather made smallholding work impossible it was not unknown for young and old to attend lessons. Schooling provided by the lead companies was not entirely free but was pitched at a modest sum which the miners could almost afford.

Recreational pursuits were varied, with religious beliefs and the Temperance Movement controlling the drinking habits of some men. Smoking was popular and broken clay pipes can still be found around the old mine sites. Recitation, yarns and singing around the fireside were regular community activities and playing a musical instrument and being a member of the local band carried some prestige. On special occasions a dance might be organised in a hall with everyone bringing some fuel along for the night. There were other events like wrestling competitions, the village fair and travelling entertainers. Sports like football, cricket and quoits had a strong following so there were teams in many villages.

At home, reading by candlelight or oil lamp was much enjoyed and all over the lead mining districts the traditional games of marbles, dominoes and pitch and toss were played.

The decline of the Lead Industry

AT the beginning of the 19th century the British lead mining industry was in its heyday. German and Spanish mines were in decline and this meant that there was little competition. Borrowing machine technology from the Germans, the British were soon the leading world producer. When wars came the price of lead was always high because lead was needed for the manufacture of ammunition. In the 1890s, however, the price of lead was reaching a low point of £9.50 a ton. Forty to fifty years earlier the price had been £21 a ton and the industry had boomed. A rapid decline set in with no attention being paid to the miner who had no union to appeal to for united action. As all of the larger veins of high grade ore had been worked it was now costly to extract ore from the smaller veins or to accept low grade ore in great quantity. There was still no cheap method of boring or tunnelling to discover extra veins.

The demand for lead tailed off as it was substituted by zinc, iron and other materials. To survive, the companies had to invest elsewhere and many found it lucrative to switch to the Spanish and German mines. The Spanish workforce could be paid less than the British and the German lead-bearing veins were more accessible than those in Britain. With the withdrawal of the companies the lead mining tradition was extinguished.

With the lead mining industry wiped out many miners emigrated to foreign mines and others joined the workforce of the nearest thriving industry. There was coal mining in Northumberland and Durham and iron and textiles in Yorkshire and Lancashire. Some men, being reluctant to leave their land and homes, turned back to farming as the company chapels, schools and mine buildings became disused and dilapidated. In recent years there has been some mining in the Dales for fluorspar which involved the opening of some disused lead mines but lead mining itself remains a sad ghost of the past.

Killhope wheel in the winter of 1983.   (S.D. Goodhart)

Lead Mining Today

IN all forms of modern mining two major issues loom large. These are firstly the need to conserve the natural resources of our world and secondly the need to avoid pollution. Steps are taken to reduce noise, dust and heavy traffic and efforts are made to reduce the contamination of our environment. Lorries are often hosed down before they leave the site to prevent lots of dirt being spread about. Two of the most important improvements are that the waste heaps which scar the landscape are hidden and that after a major mining programme the land may be reclaimed.

Mine workings which belonged to the old British lead industry are easy to spot because no effort was made to hide their presence. Unlike the concrete monstrosities which accompany modern mines, they were built in local stone and thus blend into the background of the countryside.

Today the country that produces most lead is the United States of America. Production figures constantly change and up to date facts are published in the current U.S. Department of the Interior Year Book. We can be certain however that since 1969, when the U.S.A. produced only 14% of world supplies, the output has increased until it now heads the production tables providing up to one quarter of all world lead. Other producers such as Australia, USSR, Canada, Mexico and Peru have fallen well behind the U.S. output.

There is no doubt that the production of lead will continue because of its versatility and special qualities which are as important in the Nuclear Age as they were in bygone years. Not only do we see lead derivatives in common consumer goods like crystal glass (lead oxide), storage batteries (lead dioxide) and rust preventive paint (red lead, trilead tetroxide), but lead is invaluable for radioactive protection and new techniques like lead 210 mineral dating. Whether in the U.S.A., Australia or other lands the immediate future of the lead mining industry is assured.

Lead Pollution today — A modern issue

The main social and moral issues in the old days of lead mining were the working conditions and health risks which the miner suffered in return for meagre pay. In the Western world, miners are comparatively well paid and modern mining methods have greatly reduced physical dangers.

Perhaps the most shocking problem associated with lead today is the amount which finds its way into the atmosphere. In Great Britain, many recent studies have proved that the level of lead in the air in parts of our cities exceeds the level considered safe by the E.E.C. Most of the lead in the air comes from vehicle exhaust fumes and it is absorbed into the body as we breathe. Lead is also absorbed into the body if children chew leaded paint or old water pipes. Water in these pipes is perfectly safe because oxygen from the water (H_2O) forms a layer of lead oxide on the inside of the pipes preventing the lead from coming into direct contact with the water.

Once absorbed into the body lead can have a serious effect. The consequences for children are particularly dangerous and they can become sluggish, irritable and aggressive if they are in contact with too much lead pollution. In severe cases intellectual development is impaired and other problems associated with lead poisoning are calcium deficiency and heart, kidney, blood and breathing troubles.

It is ironic that the lead miner is now less at risk from lead than many other members of society. In the United States lead-free petrol has been introduced so that exhaust fumes are less dangerous. In Great Britain action groups, with headquarters in London, are campaigning for lead-free petrol and it is likely that their schemes will eventually be introduced.

Appendix

Armchair Map Reading

Lots of interesting clues to the lead mining industry can be gleaned through browsing with a few Ordnance Survey Maps to find features and placenames. These are a lasting reminder of the old industry. Listed below are some suggestions for using the maps. Grid references and map sheet numbers refer to the new 1: 50 000 Landranger Series of Great Britain.

Northern Dales

Allendale/Cumbria	Map Sheet	Features/Placename	Grid No.
1. Carrshield	87	Temperance Farm	803473
2. Allendale Town	87	Course of old mill flue	814515
3. Carrier Way	87	Halleywell- Dukesfield smelt mill	933500
Route: Halleywell-Riddlehamhope Fell-Slaley Forest-Dukesfield			
4. Nenthead	87	Old Mill	786432
5. Nenthead	87	Chimney	794429
6. Nenthead	87	Reservoir – water to power the crushing machine	787432

Weardale

7. Moss Moor	87	Wellheads Hush	823403
8. Nunnery Hill	87	Dowgang Hush	775430
9. Killhope Museum, mine and picnic site	87	Lodging Shop, dressing floor and crushing wheel	827429

Teesdale

10. Mickleton	92	Bail Hill	970236
11. Copley	92	Lead Mill Chimney	086249
12. Carrier Way	92	Eggleston-Copley	065244

Possible routes: a. Eggleston, Langleydale Common, Copley
                 b. Hill Top, Woolly Hill, Langleydale Common, Copley

Nenthead: Exploring a mine site under instruction. The author is addressing the party. *(S.D. Goodhart)*

Southern Dales

Swaledale	Map Sheet	Feature/Placename	Grid No.
13. Gunnerside	92	North Hush	935014
14. Gunnerside	92	Disused Mines	936008
15. Gunnerside Carrier Way	92	Carrier Way to Arkengarthdale Mill	
Route: Blakethwaitye -Level House (964014)- Mill Bottom			
16. Gunnerside	92	Silver Hill (Name of old level)	934013
17. Arkengarthdale	92	Leading Stead	955063
18. Ellerton Moor	99	Old Stork Vein	066958

Wensleydale/Swaledale

19. Preston Moor	99	Old pits, old flue chimney	065940

Wharfedale

20. Grassington Moor	98	Mining area (moor)	035684
21. Grassington Moor	98	Chimney	029666
22. Grassington Moor	98	Mineworkings	025673

Nidderdale

23. Greenhow Hill	99	Disused Mines	109633
24. Greenhow Hill	99	Disused Mines	113649

Kit List

Any good books on rambling will give a detailed resume of suitable equipment for the walker. The check list below is meant as a general guide but for underground exploration your instructor will give you all the advice you need.

Wear
Boots
Thick socks, usually woollen
A watch
Warm trousers or walking breeches
Shirt
T-shirt or vest
Pullover
Anorak

Carry
Food (and spare food)
Drink (including a hot drink or facilities for making one)
Whistle (Know emergency procedure — 6 blasts per minute, rest for a minute and repeat is the international distress code)
A comprehensive first-aid kit (the lead mining dales are isolated)

Mineral collecting: searching for fluorspar crystals near a spoil heap.   *(S.D. Goodhart)*

Kagoule, waterproof trousers
Spare clothing, a jumper is essential
Torch and spares for it
Toilet requisites
Guide books, maps, compass, map case
Geological hammer, reference books, note pads
Camera
Money

Minerals and Rocks to look out for

Galena	Shiny lead ore, occurs in thin veins in limestone and gritstone. Exposed to the air the ore becomes a dull grey.
Limestone	Grey rock of calcium carbonate, often containing fossils.
Sandstone	Consolidated grains of sand with grains visible.
Shale	A soft, layered mudstone.
Calcite	White or grey with six-sided crystals, sometimes called nail head spar or paper spar or dog tooth.
Quartz	Similar appearance to calcite but you cannot scratch its surface with a coin.
Fluorspar	A light purple mineral used to make ornaments because of its colour. Often called Blue john or Derbyshire spar. Used as a flux in steel making.
Iron Pyrites	Fools gold.
Zinc Sulphide	Sometimes called Black Jack because of its black crystals. It is an ore of zinc.
Barytes	A light coloured mineral with a crystaline look. 'Heavy spar'.
Siderite	A type of magnetic ore.
Copper	Rocks containing copper have a greenish tinge.
Iron	Rocks containing iron have a reddish tinge and appear rusty.

Spoil heaps were the mounds made up of deads, the waste that was brought out of the mine. If you can find an old spoil heap and rummage around it, perhaps opening a few stones with a small hammer, then you may find lead ore or any of the above rocks and minerals. All of the minerals and rocks above are usually found in conjunction with galena.

Reading Selection

1. Clough, Robert Taylor *The Lead Smelting Mills of the Yorkshire Dales and Northern Pennines*. A study of the industrial archaeology of the lead mill, including photographs.

2. Hunt, C.J. *The Lead Miners of the Northern Pennines in the Eighteenth and Nineteenth Centuries*. The social conditions and working conditions of the miner, in detail.

3.   McTaggart, J. *Round the Hollow Hills*. The touching story of the life of Richard Watson, who was a lead miner, poet and entertainer.

4.   Raistrick and Jennings *A history of Lead Mining in the Pennines*. A book with emphasis on technical aspects.

5.   Raistrick and Roberts *Life and Work of the Northern Lead Miner*. A history of the miner, illustrated with numerous photographs of the period.

6.   Mitchell, W.R. *Pennine Lead Miner*. The story of Eric Richardson, one of the last miners in Cumbria.

7.   Turnbull, L. *The History of Lead Mining in the North East of England*. Concentrates on the Northern Dales.

The books recommended above are available through the local lending library network.

Further Information

Most book shops have a section where local history books are for sale; many of these mention lead mining. Gift shops, craft shops and newsagents near the places of interest stock the relevant books and their range is likely to grow as the leisure and tourist industry expands. Tourist Information Centres have leaflets on local towns and villages in the Dales. These leaflets are updated regularly and can be collected from any tourist office, Durham City and Richmond being the main offices for the Northern and Southern Dales respectively.

Library sections on modern mining and industries would be a good source for those interested in environmental issues and the Friends of the Earth and the Campaign Against Lead in Petrol can supply information if they are sent a stamped addressed envelope.

In 1984 a new Lead Mining Museum opened at Killhope in Weardale and information sheets can be supplied from this source for both children and adults/teachers.

Addresses

1.   Friends of the Earth, 9, Poland Street, London W1.

2.   Campaign Against Lead in Petrol, 68, Dora Road, London, SW19.

3.   Killhope Lead Mining Museum. Enquiries and information: The County Planning Department, (Countryside Team), County Hall, Durham, DH1 5FU.   Tel: Durham 64411 ext. 2534 (Countryside Team).

Index